FAIRVIEW BRANCH
SANTA MONICA PUBLIC LIBRARY

JUL 2005

21ST CENTURY DEBATES

ENERGY RESOURCES

OUR IMPACT ON THE PLANET

EWAN McLEISH

A Harcourt Company

Austin New York

www.raintreesteckvaughn.com

21st Century Debates Series

Climate Change | Energy Resources | Genetics | Internet
Media | Rain Forests | Surveillance | Waste, Recycling, and Reuse

Copyright © 2002, text, Steck-Vaughn Company

All rights reserved. No part of this book may be reproduced or utilized in any form or by any means, electronic or mechanical, including photocopying, recording, or by any information storage and retrieval system, without permission in writing from the copyright owner. Requests for permission to make copies of any part of the work should be mailed to: Copyright Permissions, Steck-Vaughn Company, P.O. Box 26015, Austin, TX 78755.

Published by Raintree Steck-Vaughn Publishers, an imprint of the Steck-Vaughn Company

Library of Congress Cataloging-in-Publication Data

McLeish, Ewan, 1950-

Energy resources : our impact on the planet/ Ewan McLeish.

p. cm.-- (21st century debates)

Includes bibliographical references and index.

ISBN 0-7398-3178-X

1. Energy development --Environmental aspects. [1. Power resources. 2. Energy development. 3. Energy conservation.] I. Title. II. Series.

TD195.E49 M38 2001

333.79--dc21 2001019048

Printed in Italy. Bound in the United States.

1 2 3 4 5 6 7 8 9 0 LB 06 05 04 03 02

Picture acknowledgments: C.F.C.L/Image Select 12; Corbis/Roger Rossmeyer 46, 48; Dennis Day 8; Ecoscene 14 (Chinch Gryniewicz), 47 (Platt), 53 (Harber); Ole Steen Hansen 13, 50; Robert Harding 44; Panos Pictures 29 (Jeremy Hartley), 32 (Sean Sprague), 42 (Giacomo Pirozzi), cover, 43 (Sean Sprague), 59 (Marc Fend); Edward Parker 18; Popperfoto 6, 7, 10, 11, 16, 20, 23, 24, 30, 34, 37, 49, 57; Still Pictures 4 (Micael Wiking), 9 (Ch. Zuber), 11 (Chris Martin), 17 (Mark Edwards), 27 (Jim Wark), 28 (Mark Edwards), 31 (John Maier), 33 (Mark Edwards), 36 (Mark Edwards), 39 (Klein/Hubert), 40 (Klein/Hubert), 56 (Sabine Vielmo), 58 (David Hoffman); The Stock Market/Lester Lefkowitz cover, 52; White-Thomson Publishing 55.

Cover: foreground picture shows: the control-room of a nuclear power station; background picture shows: an electricity-generating wind farm.

THE POWER THAT DRIVES THE EARTH

Staying Alive

It is a basic law of nature that all living things need energy to stay alive. The only difference lies in the source of that energy. For example, plants get energy from the Sun when their green leaves soak up the sunlight. They then use this energy to convert carbon dioxide and water into nourishment. Herbivorous animals obtain their energy from the plants they eat, while carnivorous animals obtain theirs by feeding on the plant eaters —and so on. A food chain like this can also be called an energy chain.

Imagine the amount of energy needed to keep all these lights burning in Hong Kong at night.

CONTENTS

Like everything else that is living, humans also need energy to stay alive, but with one big difference. We use energy not only to keep our bodies going but also to heat our homes, cook our food, provide transportation, and power our industries. Humans use a huge amount of energy: in Europe and Japan the total amount of energy used works out to about 35,000 – 44,000 kilowatt hours per person per year. This is the equivalent of every person in these countries running two large electric heaters all the time. In the United States, the rate of energy consumption is nearly double that amount. In the past twenty years, total global energy use has risen by almost 50 percent and it is still rising.

A price to pay

Humanity has become an energy-rich species. For much of our history, energy has been cheap, plentiful, and easily obtainable, but now there is a price to pay. All energy comes at a high cost, not just in dollars but in damage to human lives and to the environment. The dangers include radioactive waste, acid rain, the destruction of forests, and even flooding. A possible change in our global climate is the biggest hazard we may be facing.

Striking a balance

To safeguard the future, it is important to strike the right balance among our energy needs and ensuring future supplies without damaging the environment. This will not be easy. To achieve it, we must change both how we think about energy and how we use it.

This book will help you to examine these issues and take part in a vital debate about our energy needs and how they will be met in the twenty-first century.

FACT

25% of the world's people consume over 75% of the world's energy.

VIEWPOINTS

"Stop thinking your grandchildren will be OK, no matter how wasteful or destructive you may be. That is really mean and stupid."
Kurt Vonnegut, writer, in an address to politicians

"There are no environmental ... or resource problems that can't be solved by more economic growth, better management and better technology."
Planetary Management Worldview

These runners need energy. They get it from food rich in carbohydrates.

What is energy?

It is difficult to give an exact definition of energy. It is better to describe it as "the capacity to do work." Without it, there would be no heating, lighting, running water, transportation, or industry. Everything would come to a halt.

Frankenstein energy?

"Energy can neither be created nor destroyed." At first glance, this statement may seem contradictory. However, power plants don't create energy, they supply it in a particular form such as electricity. Sometimes, energy must be converted from one form to another before it can be used. Some forms, such as heat or electricity, are of more immediate use to us than others. For example, the chemical energy in food is not much use to us until it is released in our bodies during a process called metabolism. Then it can be used to move our muscles, heat our bodies, or make new cells.

Energy cannot be destroyed but it can be "lost." Unwanted or unused energy is simply converted into a less useful form, such as heat or noise. Once changed, it is almost impossible to convert it back to a more useful form of energy. In fact every time work is done, an energy conversion occurs and some energy is lost. Most of our machines (including our own bodies) are not very good at converting energy. As we will see later, this is why we are so good at wasting it!

FACT

An average-size power plant run on fossil fuels or nuclear energy, producing about 500 megawatts of power. That is 500 thousand kilowatts or 500 million watts.

Energy versus power

Energy and power are not the same thing. Energy is what drives things and makes them work. It is measured in joules or calories. Power is the rate at which that work is done and this is measured in watts. The more powerful a machine is, the faster it can perform a particular task. Since joules and watts are very small units, we normally refer to kilojoules (1,000 joules) or kilowatts (1,000 watts). For example, the power of an electric coffee pot is around 2.5 kilowatts (2,500 watts).

Understanding the difference between energy and power is crucial to understanding the issues surrounding how we obtain, use, and waste energy. It is also crucial to understanding why we need to get more power from less energy than we do at present.

These bullet trains in Tokyo are superbly streamlined to reduce drag and make them more energy-efficient.

VIEWPOINT

"It is clear that a low energy path is the best way toward a sustainable future."
Gro Harlem Brundtland, "Our Common Future"

A brief history of energy

Imagine that it is 100,000 years ago. This is about the time that human beings discovered how to make and use fire. They had known of its existence for thousands of years; they had even used it sometimes when a chance bolt of lightning ignited a tree or dry grass. But now they would produce it whenever they wanted to and use it to keep warm at night, scare off wild animals, cook their food and survive in colder climates. The discovery of fire as a useful energy source was a huge step forward for humanity.

Steam power

As the human race developed, it discovered other sources of energy. For thousands of years, people harnessed the energy of water and wind to drive machinery. By the eighteenth century, the more developed nations had invented the technology to produce steam under pressure. This powered machinery more efficiently than the old water- and windmills. Huge steel and cotton mills began to dominate the landscape. This development is known as the Industrial Revolution.

The energy from moving water has been used for thousands of years to power machinery. These waterwheels in Syria are more than 500 years old.

Just over a hundred years ago, the discovery of oil led to the inventions of the internal combustion and diesel engines. These inventions made possible the development of even more powerful machines and a new form of transportation which would change the face of the world—the automobile.

The secret of the atom

Sixty years ago, scientists working on a secret project in the United States solved the ultimate riddle of the time—how to split an atom and release the powerful energy contained in its nucleus.

The ultimate power source? Splitting the atom created a new kind of energy that would change people's lives forever.

The result of this discovery was the building of the atomic bomb, used to destroy two Japanese cities during World War II. That same energy has also been harnessed for more peaceful purposes such as radiation to treat cancers and as nuclear power to satisfy the world's increasing demand for fuel.

Back to the future

Today we must look forward and consider what sources of energy will sustain us in the future. Some new sources of energy may not be discovered, while others might involve using familiar sources in new ways. Sometimes we have to look back before we can look forward to the future.

VIEWPOINT

"What is clear is that the energy future will be very different from the energy past."
J. G. Soussan, lecturer and researcher in energy and resources

A growing problem

During the past twenty-five years, the world's energy needs have grown at the rate of about 2 percent each year. This increase shows little sign of slowing as the developing countries of Asia, Africa, and South America begin to experience their own industrial revolutions.

Today, our demand for energy is affecting the whole globe. Most scientists are now certain that some of the recent changes in global climate, such as higher temperatures and more extreme types of weather, are the result of increased levels in the atmosphere of so-called greenhouse gases such as carbon dioxide. Greenhouse gases are produced when fossil fuels (coal, oil, and gas) are burned. In addition, many of the world's cities are veiled in a continuous fog of exhaust fumes and harmful gases produced by industrial processes, while entire forests stand silent and bare, damaged by acid rainfall. In poorer developing countries, too, forests are steadily diminishing as the hunt for firewood becomes more desperate.

A heavy curtain of smog hangs over Chile's capital city, Santiago. The pollution is caused by a mixture of vehicle fumes and industrial emissions.

Don't breathe the air! A Thai traffic policeman adjusts his face mask as he goes on patrol in Bangkok. Traffic police, pedestrians, and street sellers often wear masks to protect themselves against toxic fumes produced by the city's cars and factories.

Getting it right

There is plenty of evidence to show that the Earth has suffered much damage as a result of our energy use. We don't yet know what hidden damage remains to be uncovered or what effect this will have on our planet's future. Much will depend on what happens to the way we produce and use energy in the next 50 years or so—that is, within your lifetime.

Scientists are currently working on the development of new fuels that are less harmful to the environment. They are also redeveloping old energy sources—wind, water, and sunlight.

Despite our enormous energy consumption, much of the world's population still receives little or no benefit from it. For example, it is estimated that one person in Burkina Faso, West Africa, consumes less than one percent of the commercial or generated energy used by one person living in Canada.

It is issues like these, as well as environmental concerns, that we must address in order to get a balanced view of the world's energy needs in the twenty-first century.

DEBATE

Why is it that we are so reliant on energy resources? Are we, by nature, an "energy-rich" species? Is it inevitable that we continue this energy dependence, or are there other ways of living in which energy is seen as a privilege, to be used carefully, rather than as a commodity to be used at will?

FOSSIL FUELS

The Great Carbon Show

About 75 percent of the world's energy comes from fossil fuels: coal, oil, and gas. These sources formed from the preserved remains of trees, plants, and microscopic marine animals that lived millions of years ago. When they were alive, the trees and plants absorbed carbon dioxide from the atmosphere, while the sea creatures fed on marine plants and absorbed the carbon from the plants into their bodies. When they died, many trees, plants, and animals sank to the bottoms of swamps, rivers, lakes, and seas, and were covered with sediment which preserved them. Gradually the remains, which consisted mainly of carbon, changed into coal, oil, and gas—the future fossil fuels.

A North Sea oil rig burns off waste gas. Oil and gas still make up the largest source of energy in the world.

Full circle

When fossil fuels burn, the carbon they are made from combines with oxygen to form carbon dioxide (CO_2). In this way, the carbon that was once absorbed by plants and animals millions of years ago is returned to the atmosphere. However, the fossil fuel that took nature about a million years to produce is being burned today in a year. Two inescapable conclusions arise from this fact: one is that fossil fuels are being used much more quickly than it took for them to form (see page 14), the other is that carbon dioxide levels must be rising (see pages 18–19).

Super-energy!

Compared to many other sources of energy, fossil fuels are easy to obtain, convenient to use and, in some cases, still in plentiful supply. Perhaps more important, fossil fuels are a highly concentrated source of energy. This is why they make good fuels—they deliver large amounts of energy fast. As we shall see next, this has been important in the way different countries have developed. For much of the last century, those who held the biggest energy resources also held the most power.

FACT

The United States has only 4% of the world's oil reserves but uses nearly 30% of the oil extracted worldwide each year. 66% of this oil is used to transport goods.

A coal-fired oil refinery in Denmark. Refineries tend to be near the coast so that they can easily receive crude oil from tankers. However, this increases the likelihood of coastal oil pollution.

VIEWPOINTS

"Would Britain let BP [British Petrolium] drill oil in the Lake District?"
Randall Snodgrass, WWF [US] on BP's proposals to drill for oil near caribou calving grounds in Alaska

"This is all built with oil money (schools, clinics, fire engines). We have our own company that owns oil land. Many local people here are shareholders. Without oil this place will go down."
George Tagorook, Inupiat [Eskimo] and deputy mayor, Kaktovik, Alaska

Going, going, gone?

Fossil fuels cannot last forever. But no one really knows how much oil, coal, and gas remains or how long supplies will last. As known reserves are used up new ones are still being discovered, meaning that the "horizon" keeps moving farther away.

New exploration and extraction techniques also mean that fossil fuel reserves that were once unobtainable have now become accessible. At the same time, improved technologies make extracting from what were once thought to be exhausted reserves possible.

Present estimates suggest that oil and gas reserves could last for another 50 to 100 years, while coal could last for several centuries. These estimates depend on how much fossil fuel is used compared to other types of fuel. It will also depend on whether the world's nations allow new reserves of fossil fuel to be exploited, such as those recently discovered beneath the North Atlantic and the Arctic Oceans.

These trucks are equipped with vibrating plates able to detect the location of buried oil reserves in the Sahara Desert.

The oil weapon

Some countries have more fossil fuel reserves than others. China, India, and Russia have huge deposits of coal while Russia also has massive reserves of natural gas. South American countries, such as Venezuela, have large oil fields but most of the world's oil is found in the Middle East in countries such as Saudi Arabia and Kuwait.

Most of the oil-producing countries belong to OPEC (Organization of Petroleum Exporting Countries) which

This is an open cast mine, where the coal is obtained directly from the surface. Open cast mines, such as this one in India, can cause large-scale damage to the environment.

allows them to dictate both the supply and price of oil to non-producing countries. This may not always be a bad thing since high prices may reduce consumption, but it can also put pressure on poor countries to develop other sources of energy, such as nuclear power.

Equally serious is the fact that oil-rich countries may face invasion by other countries wishing to take over their reserves. It was for this reason that Iraq invaded Kuwait in 1990. To protect Kuwait's oil reserves, Great Britain, the United States, and many other countries launched an attack on Iraq. Although Kuwait was freed from Iraqi control as a result, thousands of Iraqi soldiers lost their lives and the Iraqi people have suffered hardship ever since because of trade sanctions placed on them by the rest of the world.

FACT

The 13 countries that make up the Organization of Petroleum Exporting Countries (OPEC) control 67% of the known oil reserves.

FACT

Lakes and streams in Europe's highlands may not become acid-free in the near future even though power plants are cutting their sulfur emissions by 90%. This is because underground reservoirs of acidic water built up over many years may go on pumping acid into streams for decades to come.

A cloud on the horizon

When coal and oil are burned in power plants, industrial furnaces, or car engines, they produce, emissions (gases) that contain sulfur and nitrogen. When these gases combine with water vapor, sunlight, and oxygen high in the atmosphere, a mixture of acids is produced that can travel for hundreds of miles around the world until it falls as acid rain or snow. The acid attacks buildings, monuments, and even stained-glass windows: when it falls into lakes and rivers, plants and animals are harmed. When it falls on land, nutrients are washed out of the soil, making plant life more vulnerable to disease, frost, and drought.

The good news

In recent years, the amount of damage from acid rain in the West has fallen. This is because Europe and the U.S. now set strict limits on acidic emissions from industries. A system of high-temperature combustion and filters removes much of the sulfur from power-plant emissions. In addition, all new cars are fitted with catalytic converters which take out many of the pollutants, such as nitrogen dioxide and carbon monoxide, from their exhausts.

The bad news

Unfortunately, catalytic converters become less efficient with age and do not work

Because of the amount of pollution pouring from the city's 10,000 factories, life in Changoing, China, is difficult.

Testing the rain for acidity in Sweden. The country spends over $15 million a year adding lime to lakes in order to reduce acid levels. Ten percent of Sweden's acid rain comes from Britain.

properly until the car engine has heated up sufficiently. Since many car trips are short ones in towns or cities, levels of acidic gases have not been reduced as much as expected. In addition, the sheer number of cars and other vehicles on the road continues to increase (see pages 34–37). In industries, although sulfur emissions have been reduced dramatically, nitrogen emissions have hardly changed.

The really bad news

By the year 2010, sulfur emissions in Asia are set to triple from 1990 levels. As developing countries such as Indonesia and China become more industrialized, their energy needs are skyrocketing. The most common fuel used in their power plants is coal high in sulfur.

Unless the developing countries get more help from the richer nations to create cleaner technologies, acid rain, and other pollutants will continue to damage natural environments.

VIEWPOINTS

"The use of the best available pollution control technologies could cut acid deposition ("rain") levels in half by 2020 in Asia, even though energy use will triple. But the cost is high—$90 billion every year."
World Bank

"Reduction strategies [for acid rain] are a cheap insurance policy compared with the vast amount of potential damage they seek to avoid."
Gro Harlem Brundtland, "Our Common Future"

VIEWPOINTS

"Does the barren landscape of Mars represent our near future? Reducing carbon dioxide emissions may help. However, it may be that, in environmental terms, we have already pushed a very large rock off a very steep hill."
Chris Clayton, engineer.

"Scientists who claim [that] the evidence supports global warming are guilty of "'scientific cleansing'" [i.e. distorting evidence that does not support their view]."
Global Climate Coalition [a body representing oil and vehicle companies]

Global warning

All fossil fuels produce carbon dioxide when they burn. Unlike pollutants such as sulfur and nitrogen oxides, carbon dioxide cannot be easily removed by filters or high-temperature combustion. Carbon dioxide is the main (but not the only) cause of global warming. As the level of carbon dioxide in the atmosphere increases, less of the Earth's heat is able to escape into space and is reflected back toward the Earth, causing the planet's surface to warm up.

CO_2 on the rise

There is no doubt that carbon dioxide levels are rising. Concentrations of CO_2 in the atmosphere

The effects of global warming, caused mainly by rising levels of CO_2, may mean worsening droughts in African countries such as Mali.

have increased by over a third since Europe's Industrial Revolution 200 years ago. As a result, global temperatures have risen by one degree fahrenheit over the last century. While this may not sound like much, the effects could be far greater than we imagine. It is also thought that the present rate of warming has increased to about two or three degrees per century.

A matter of opinion?

Most scientists agree that some global warming is now happening, but few agree on what the likely outcome will be. Some believe that the effects are being felt already—signs include warmer air and sea temperatures, a rise in sea level, and erratic weather patterns. Some experts think that ocean currents such as the Gulf Stream will be affected, causing dramatic climate and weather changes worldwide. Sea-level changes could cause low-lying countries such as Bangladesh to suffer increased flooding, and many Pacific islands could disappear. It is also possible that tropical forests may die from "heat stress" caused by increased global temperatures. Others believe that these claims are highly exaggerated and that any changes to the environment are due to natural variations, such as the change in weather patterns known as El Niño. This occurs naturally every two to seven years when winds from the coast of Peru push warm surface water offshore, allowing cold, deep water to rise to the surface. This affects ocean circulation which, in turn, creates short-term climate changes worldwide.

We do not really know how damaging the effects of increased carbon dioxide levels and global warming may be. Some places may actually benefit from a warmer climate, at least for a time. However, it is more likely that the overall effects will be very damaging.

FACT

The checkerspot butterfly of North America is the first convincing evidence that the geographical range of an animal species has shifted in response to climate change. Butterfly colonies on the southern limit of its range have failed to survive, while new colonies have formed on the northern limit.

DEBATE

How do we know whom to believe when we hear different claims from experts about issues such as global warming? Is scientific evidence always impartial, or is it likely to be influenced by "vested interests" (i.e. people or corporations supporting a specific view to gain something)?

NUCLEAR POWER

Releasing the Genie

FACT

Scientists have discovered that one of the main ingredients of the nuclear fuel, MOX (mixed oxide), can react with water and therefore spread in underground water systems.

Nuclear power has been available since the 1950s. At that time, scientists and politicians predicted that the energy it produced would be "too cheap to meter." It was also thought that by the end of the century most of the world's electricity would be produced in nuclear power plants. In fact, only about 17 percent of today's world supply is nuclear. Although nuclear power still has its champions as an alternative to fossil fuels, it is likely that the twenty-first century will see the end of nuclear power. We may have to wait tens of thousands of years, however, to see the end of its legacy—nuclear waste.

Anti-nuclear protestors make their feelings clear about the dangers of nuclear power.

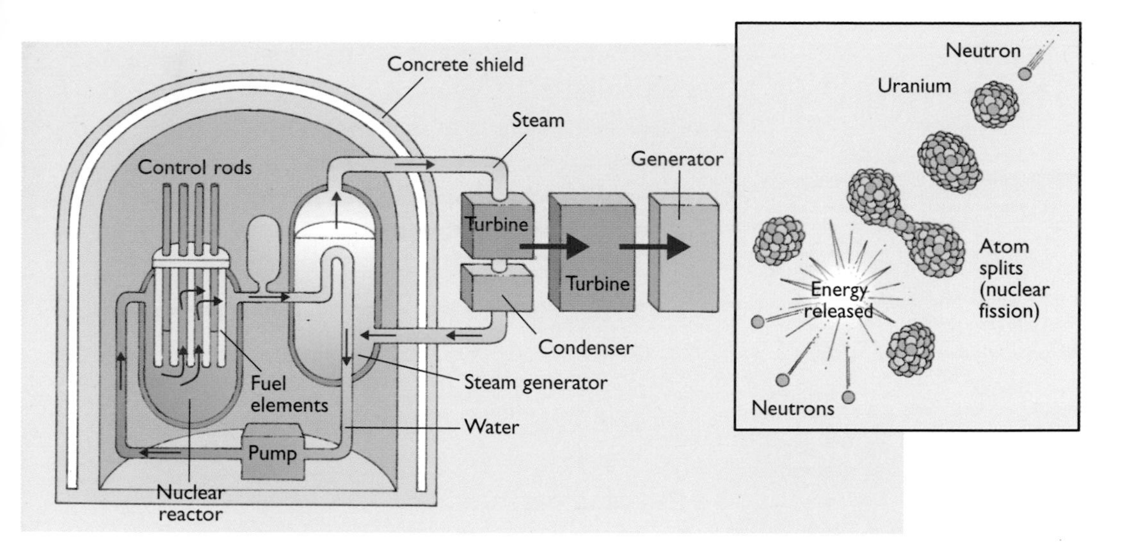

In a nuclear power plant, uranium atoms are bombarded with neutrons. The atoms split and release large amounts of energy and more neutrons which then smash more uranium atoms in a chain reaction. The energy (heat) produces steam which drives turbines.

How does nuclear power work?

All atoms are made up of a central nucleus and orbiting electrons. The nucleus itself is made up of smaller particles—protons and neutrons—held together by a nuclear force. Certain types of atoms, called isotopes (e.g. uranium-235 or plutonium) are large and unstable. When they are bombarded by neutrons, these isotopes can be split into smaller atoms (e.g. barium and krypton) releasing huge amounts of energy, and more neutrons. These neutrons collide with other uranium atoms, starting a chain reaction that results in the release of huge amounts of energy in the form of heat. The heat is used to produce steam to drive turbines and generate electricity.

No global warming, no acid rain

Energy from nuclear power produces fewer air pollutants than burning fossil fuels. Like fossil fuels, nuclear reactors can deliver large quantities of energy quickly. Nuclear reactors are built to high safety standards, equipped with so-called "multiple back-up systems" that are designed to prevent accidents from happening. Most reactors are very safe most of the time.

VIEWPOINTS

"At the present time, the atom is the only source of energy that can produce electricity on a comparable basis to fossil fuels—without the air pollution."
Nuclear power industry promotion

"Today, if you had tons of plutonium to offer for free, no one would take it."
Klaus Janberg, German nuclear waste company

VIEWPOINTS

"When I went into the schools in Byelorussia (Ukraine), I learned that the first-graders have never been in the forest because the trees are so contaminated. When children want to see what nature used to be like, they go into a little courtyard inside the building, and the teacher says, "This is a bird and this is a tree," and they are plastic. Isn't that sad?"
Olga Korbut, Ukrainian former gymnast, on Chernobyl

"The likelihood of a major nuclear accident is one every 100,000 years of operating time."
Much quoted figure by the nuclear industry

When things go wrong

Unfortunately, some nuclear reactor designs are better than others and, however good they are, human error can cancel out even the best safety systems. In March 1979, one of the nuclear reactors at the Three Mile Island power plant in Pennsylvania, lost some of the coolant water that keeps the reactor core from overheating. Some of the core melted and large amounts of radioactivity escaped into the atmosphere. About 100,000 people had to be evacuated from the area.

History repeating

Seven years later, at the nuclear power plant in Chernobyl, Ukraine, engineers turned off most of the reactor's safety and warning systems in order to conduct a safety experiment. A series of explosions ripped through the plant, blowing off the massive concrete roof of the reactor building and flinging radioactive dust high into the atmosphere. The dust eventually circled the Earth.

The cost

Serious damage to the health of thousands of people at Three Mile Island was probably avoided by no more than 30 minutes. At Chernobyl, 31 people died soon after the accident as a result of massive radiation sickness and, of the hundreds of others who needed treatment, several died. Over a quarter of a million people were evacuated from the area and many will never be able to return because of high radiation levels. When rain contaminated with radiation fell in western and northern Europe, farm animals were slaughtered because they were too radioactive to be used for meat.

No one knows what the long-term effects of the disaster at Chernobyl will be. Over half a million people were exposed to significant levels of radioactivity, and it is still too early to say how many deaths from cancer or thyroid tumors will eventually occur.

Sleeping giants?

Some experts think that another nuclear accident is inevitable. Some of the older nuclear power plants in the countries of the former Soviet Union (USSR) are especially risky and many people in the West believe these should be made safe and shut down. There are over 400 nuclear power plants throughout the world, few of which have experienced problems since Chernobyl. However, the possibility of an accident is not the only hazard of nuclear power.

Safety first? People form a line to receive a medical examination in Mito, Japan, in October 1999, following a uranium accident which sent radiation levels soaring at a nearby nuclear plant. At least 49 people were exposed to radiation and 320,000 people were forced to stay indoors.

FACT

In July 1999, two consignments of reprocessed fuel rods destined for Japan were refused when it was discovered that safety records at the reprocessing plant at Sellafield, England, had been falsified.

The environmental legacy of nuclear power

All energy generation produces some kind of waste, and low-level radioactive waste is produced during all parts of the nuclear power production process. This waste has to be stored safely for several decades before it is reasonably safe; it is usually packed into steel drums and buried in "landfill" sites.

Many environmentalists believe that this is unsafe. They claim that some radioactivity will eventually leak out and may contaminate underground water supplies. They argue that the waste should be kept above ground at the site where it is produced.

Nuclear plants also produce large quantities of waste water, used in the cooling part of the process. Most of this water is discharged into the sea where it might contaminate beaches or enter the food chain.

The quarter-million year waiting list

High-level nuclear waste, mainly from the reactors themselves as well as from nuclear weapons, gives off large amounts of radiation and has to be stored for many thousands of years. There is no real agreement on a safe way to store the waste.

A steel case of mixed plutonium-uranium oxide (MOX) is delivered to Japan from a British registered ship in October 1999. Earlier deliveries had to be returned for safety reasons.

Some scientists believe that it should be buried deep underground after first being "re-processed" to remove very long-lived isotopes such as plutonium. The highly radioactive remainder would be trapped in glass and sealed in huge metal canisters before being stored above ground or buried. Further suggestions for disposal include burying the waste deep beneath the polar ice-caps or in deep ocean-floor mud, and even shooting it far into space.

Does anyone want an old nuclear power plant?

The operating life of a nuclear power plant is about 40 to 50 years, but usually less. Old nuclear power plants cannot simply be demolished—they must be "decommissioned." This involves removing all the high-level radioactive material contained inside and then finding a way to keep the reactor isolated and safe for thousands of years, such as entombing it in concrete. Between 2000 and 2012, over 200 large commercial reactors will need to be decommissioned, but as yet no safe system for doing so has been agreed.

The future of nuclear power

Very few countries now see a future for nuclear power. Even if there were no environmental or safety concerns, the financial costs are now known to be two or three times higher than many other forms of power generation.

In June 2000, Germany announced that it would close down all its nuclear power plants as soon as possible (a process that may take as long as 30 years). A small number of countries, with few energy sources of their own, are likely to continue with their nuclear power programs but, for most countries, the idea of cheap, safe, and unlimited energy supplies has proved to be no more than a myth.

DEBATE

As technologies improve, do you think the chances of another major nuclear accident are less, rather than more, likely? Is the disposal of nuclear waste *our* problem or will future generations find ways to make it safe?

WATER POWER

Making Water Work

A high mountain river, or water that is contained behind a massive wall or dam, contains so-called "potential energy." When the water is released, the energy is liberated and can be used to drive turbines and generate electricity. This is called hydroelectric power and it supplies about 25 percent of the world's electricity and 5 percent of its total energy consumption. Norway's mountainous landscape and numerous rivers enable it to obtain 95 percent of its electricity in this way.

VIEWPOINTS

"The threat from the Tehri Dam (in India) is made worse by its being constructed in a seismically active zone ... if a future quake breached the dam, tens of thousands would die."
Anil Agarwal, Center for Science and the Environment, India

"The (Tehri) Dam will bring better schools, better health care, a better life to thousands of people; many villages will have electricity for the first time; how can that be bad?"
Local engineer

Too good to be true?

Hydroelectric power avoids many of the problems connected with fossil fuels and nuclear power. Hydroelectric plants rarely need to be shut down, they have fairly low operating and maintenance costs and, most important, they produce no carbon dioxide or other forms of pollution. In addition, a hydroelectric plant has a much longer life span than a fossil fuel or nuclear power plant, and can help to control flooding and irrigate farmland besides supplying electricity for thousands of isolated villages. This is particularly useful in developing countries where other energy sources may be too expensive.

Spot the drawback

Unfortunately, a very large dam-building project can bring with it a number of serious disadvantages. Building the dam requires a huge input of energy and materials that can be enormously damaging to the local environment. Diverting a river may deprive large areas of farmland downstream of nutrient-rich silt, normally deposited by seasonal river floods. Most seriously, the enormous weight of water held by the dam may create faults in the reservoir basin which can set off local earthquakes and cause flooding, and damage and loss of life downstream.

The reservoir can also become a breeding ground for malaria-carrying mosquitoes while silt, carried into the reservoir from upstream, decreases the dam's efficiency and shortens its life span. Most damaging of all, flooding the land to create the reservoir in the first place destroys forests or farmland, and often displaces thousands of people.

FACT

The total amount of hydroelectric power produced per year is about the same as 500 million tons of oil, or over one-sixth of total oil production. This is about 500,000 megawatts.

Glen Canyon dam and hydroelectric plant, Arizona. Hydroelectric power does not have the disadvantages of fossil fuels but massive dams bring their own problems.

FACT

The age of building large dams in the United States is ending because construction costs are too high, there are few suitable sites left, and there is too much opposition to proposed new projects on environmental grounds. Any new large supplies of hydroelectric power will be imported from Canada, which gets 70% of its electricity from hydro-power.

Empty promises

The Senegal is the second longest river in West Africa, passing through Guinea, Senegal, Mali, and Mauritania. An ambitious project to build two dams along the river to generate electricity, improve fisheries, create freshwater supplies, and improve irrigation and agriculture began in the 1980s. The large, hydroelectric Manantali Dam was completed in 1987 and operational by 1991.

Today, the project has yet to bring any major economic benefits to the region. Commercial electricity has not yet been produced and it is doubtful that the full plan, if ever completed, will fulfill, the promises made 20 years earlier.

The damage

The lake surface behind the dam covers over 174 square miles (450 sq km) of former forest land. In Mali, more than 10,000 people were relocated because their villages were flooded; fish catches fell, because people either moved away or were not skilled at fishing in deep lakes; and health risks from water-borne diseases increased.

The seasonal river flooding of their plots was also interrupted and many people stopped growing a healthy variety of vegetables and cereals and began to grow only rice. The changes in land use also created tension between cattle ranchers and farmers, worsening existing conflicts in Mauritania and Senegal.

Harvesting irrigated rice in Burkina Faso, Africa. Dams can improve irrigation by controlling water supplies, but they also disrupt natural flooding which brings valuable nutrients to farmland.

A dam provides the energy source, a micro-turbine generates the electricity and power lines carry it to a community in Nepal.

Finally, most of the electricity is intended for the region's cities rather than local communities.

Think small?

The answer to the problem of large dams seems obvious: build larger numbers of small dams. They cause less damage to the environment and can provide energy exactly where it is needed. In China today, 90,000 small dams provide 40 percent of its small towns and 30 percent of its cities with most of their electricity.

However, small dams can be less economical to build or run than large dams and suitable local sites or a skilled workforce may be difficult to locate. In developing countries, these projects are usually funded by donors in the developed countries who often prefer to put their money into large, high-profile projects rather than small, local ones. There are no easy solutions.

DEBATE

Many of the problems associated with the Manantali Dam were predictable. Would better planning make such projects more successful? Hydroelectric power allows poor countries to avoid importing fossil fuels, allowing them to spend more on other aspects of development. Does this outweigh some of the disadvantages?

ENERGY FROM PLANTS

The Fuel of the Poor

VIEWPOINT

"Because biomass energy is the fuel of the poor, it plays no part in the international energy debate. It is invisible."
United Nations

Wood still meets the main energy needs of over half the world's population, mostly in developing countries in Africa and parts of Asia.

Wood and other plant and animal material is known as "biomass." Burning biomass to cook food and heat buildings provides nearly 15 percent of the world's energy needs (36 percent in developing countries).

Biomass is a fairly clean form of energy, producing far fewer acid gases than coal or oil. Although biomass produces carbon dioxide when it burns, this replaces the CO_2 that was removed from the atmosphere when the plant was alive. As long as plants are replaced at the same rate as they are burned, biomass is "neutral" in terms of carbon dioxide production.

An Ethiopian woman carries a load of cow dung to market to sell as fuel. It would be better if this dung could be used to fertilize Ethiopia's poor soils.

In Brazil, to make charcoal, wood is burned in kilns from which air is excluded. This makes the wood lighter to transport, but it also reduces its energy value.

A big "if"

However, in most countries where biomass is an important fuel, the rate of use is much greater than the rate of replacement. The result is that large areas are stripped of trees. In Burkina Faso, West Africa, it is estimated that people have to walk for three-and-a-half hours a day to collect and carry 88-110 pounds (40–50 kilos) of wood. They therefore spend less time growing food. Another impact is that, as wood becomes scarce, people are forced to burn animal dung and crop waste which would normally be used to fertilize the soil.

People's health is also affected, especially children's, as quicker-cooking cereals replace more nutritious but longer-cooking vegetables. The lack of clean, boiled water also means that disease is more likely to spread.

The city strikes back

As more people move from rural areas into the cities, the demand for energy increases. In developing countries this usually comes in the form of charcoal obtained by burning wood in a limited air supply. Charcoal only contains about half the energy value, so twice as much wood is needed to produce the same amount of energy.

FACT

Even in oil-rich countries such as Nigeria, fuelwood still makes up over 75% of energy use. This is because much of the oil is exported to produce income for the country.

At this filling station in Brazil, alcohol is used to power cars. Although alcohol is a fairly clean fuel, its production may be damaging to the environment.

One gallon of alcohol, please!

Wood is not the only source of biomass. Ethanol (alcohol) can be made from sugarcane or sugarbeets by a process called fermentation. Ethanol is a suitable car fuel and is used in countries that have poor fossil-fuel reserves but warm climates and large amounts of agricultural land, such as Brazil. However, growing crops to fuel cars still requires fertilizers, pesticides, irrigation, harvesting, and a workforce, adding up to little overall saving over the cost of using fossil fuels. Methanol is an alternative alcohol fuel that can be made from natural gas, coal, or agricultural waste, but its production creates many of the same problems as any other use of fossil fuel.

Grow it, read it, burn it

More and more cities in Japan, western Europe, and the United States are building incinerators to burn trash instead of burying it in landfill sites. The advantage is that the incinerators provide energy to reproduce electricity or to heat buildings. However, although toxic emissions can be kept at low levels, most people living locally object to incinerators and environmentalists argue that it is far more energy efficient to compost or recycle organic waste and paper than to burn it.

FACT

Recycling paper saves 2–4 times more energy than burning it.

Waste into energy

Solid animal, human, and crop waste can be converted into gas and liquid fuels. China has over six million biogas digesters in which bacteria break down waste products to produce bio- or methane gas. Once the gas has been collected, the remaining solids are used to fertilize crops. However, existing biogas converters are often slow and unreliable.

Biomass and the future

It has been estimated that present supplies of wood and agricultural waste products could produce up to 30 percent of the world's electricity. However, growing biomass fuel uses up a lot of land. Unless the land is properly managed and regularly replanted, soil erosion and poor fertility will lead to environmental damage.

DEBATE

Biomass does not add more CO_2 to the atmosphere—but should the land be used for growing food instead? Even in developed societies, poorer people have less choice over what kind and how much fuel they can afford to use. Do people have a basic right to supplies of fuel? And if so, what does this mean for the world's energy supplies?

In Andhra Pradesh, India, a biogas digester converts animal and vegetable waste into methane.

TRANSPORTATION

VIEWPOINT

"If the price of gasoline were to increase in the United States to reflect its full costs—economic, environmental and political—consumers might demand more fuel-efficient cars."
J. H. Gibbons, environmentalist

The Curse of the Car!

In the West, our lives would be unimaginable if we could not travel to other places or rely on having goods delivered to us. These essential services rely on transportation, which uses up more than 25 percent of the world's energy consumption including 50 percent of its oil. Motor vehicles—cars, trucks, motorcycles, and buses—make up 80 percent of the total and so the way we travel, or transport goods, has a major impact on our use of energy.

For the last 30 years, the global number of motor vehicles has been rising by about 16 million per year. In spite of lower gasoline consumption and catalytic converters, air pollution is a major problem. In Mexico City, for example, the air pollution is so bad that motorists are only allowed to drive in the city on certain days.

More cars = more pollution

The present ratio of people to vehicles in China is about 8 per 1,000, and in India 7 per 1,000 – compared with more than 750 per

Traffic chaos hits the streets of Dhaka, Bangladesh. Even the large number of pedal rickshaws does little to reduce the number of motor vehicles.

1,000 in the United States. As the economies of poorer countries strengthen, the demand for cars will rise. Predictions place the number of vehicles on the world's roads by the year 2025 close to a billion.

Enter the electric car

Electric vehicles are quiet, need little maintenance and can accelerate as rapidly as most gas or diesel-powered vehicles. Most run on batteries, although some may soon run on fuel cells powered by hydrogen (see page 49). Of course, although the electric vehicles themselves produce no emissions, the same may not be true for the electricity needed to recharge their batteries (sometimes called "elsewhere pollution"). However, if solar cells or wind turbines (see pages 40–43) are used to generate this electricity, there is no such problem.

Cool cars

Modern electric cars can perform as well as most conventional cars and can travel several hundred miles before the batteries run down. Vehicles are now being developed that use super-light, extra-strong materials and can travel even farther.

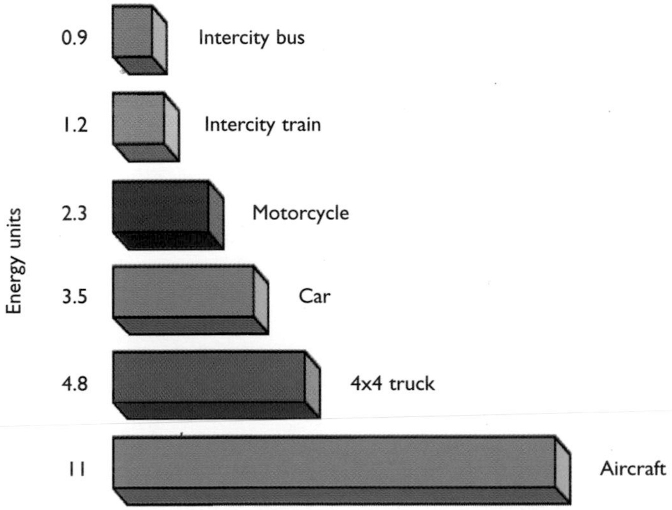

FACT

An Israeli company has produced long-range batteries that are removed entirely from the vehicles after they have run down and replaced by fresh ones. This shortens the refuelling time dramatically. The run-down batteries are then recharged and used again.

Energy units

0.9 Intercity bus

1.2 Intercity train

2.3 Motorcycle

3.5 Car

4.8 4x4 truck

11 Aircraft

A comparison of the amounts of energy units needed by different forms of transportation per person per mile (passenger miles). A car will use less fuel than a bus, but will carry far fewer passengers. So it's overall energy efficiency in passenger miles is nearly four times less than that of a bus.

FACT

In Europe, every capital city except Paris and London has a trolley network. Many of these now have computerized systems that give them priority over all other vehicles at intersections, making them economic to operate, energy efficient, fast, and safe.

Get off the road!

The change to vehicles powered mainly by non-fossil fuels will take time, perhaps from 30 to 50 years. Many environmentalists argue that we do not have that much time—they believe the only way to avoid disastrous damage to the atmosphere (as well as massive congestion on the roads) is to change our traveling habits now.

Let the train (and the bus) take the strain

We love the car—even when it can be shown without doubt that public transportation, (buses, trolleys, trains, subways) is quicker, cheaper, and safer, people are still very reluctant to give up what they see as the comfort, convenience, and independence of their own vehicles.

Cars are not always the best form of transportation. Millions of commuters drive to work every day, with only one person in a car. They often find themselves caught in traffic jams, which is not only frustrating, but adds to the total pollution load.

Buses and trains use 75 percent less energy per person than cars. Many countries now have modern and efficient public transportation systems, some of which are "integrated systems" in which

A trolley station in Amsterdam, Holland, a city in which people are encouraged to choose public transportation to get around.

the different services link up and where it is possible to use both private and public transportation together. However, in other countries, public transportation is poor, services are unreliable, and fares are high. Until the governments in these countries invest more in public transportation, people's reliance on the car will continue.

Freight fright

Most goods are transported by road rather than by trail. In Europe, 95 percent of all agricultural products, 97 percent of all foods, 75 percent of petroleum products and 98 per cent of building materials and manufactured articles are carried by road. Only solid fuels, such as coal, metal ores and fertilizers, are transported by rail in significant amounts.

With better rail links and services this could change. Goods could be transported long distances by train and then transferred onto trucks for final delivery. This would also relieve congestion on many roads.

The road freight industry has a lot of influence. Many goods delivered by road could be transported by train, causing less environmental damage.

The price of fuel

The price of oil (and therefore of gasoline) is controlled to some extent by the oil-producing countries (see pages 14–15). However, individual governments affect prices even more by the amount of tax they add to fuel. In the UK, for example, fuel tax is high because the government believes that this will reduce car and truck use and make other forms of transportation more popular as well as providing government revenue. In other countries, such as the U.S., fuel taxes and therefore fuel prices are much lower. When this book was written, the price of gasoline in the UK was four times that in the U.S.

DEBATE

Do people have the right to use whatever form of transportation they like? Most of us want cheap fuel, but should governments always do what is popular, even at the expense of the environment? Do higher fuel prices really affect people's travel habits?

RENEWABLE ENERGY

A Difficult Choice

VIEWPOINTS

"It is hardly surprising that the political will to challenge the powerful interests that benefit from the present energy economy, is absent."
J. G. Soussan, energy expert

"Renewable energy, especially from the Sun, will dominate energy production by 2050."
Shell International Petroleum

Fossil fuels will eventually run out, or be so difficult to obtain that it will no longer be worth extracting them. They are "non-renewable resources" meaning that once used, they cannot be renewed or replaced. Many people also believe that the damage they do to the environment makes it too dangerous to go on using them.

The same is true of nuclear power—uranium is ultimately a "non-renewable resource" and the dangers of continuing to use it make its future as a safe, reliable energy source doubtful.

Hydroelectric power comes from an energy source that is constantly being recycled or renewed by the water cycle. However, as we have seen, large dam projects can cause great damage to the environment as well as other problems. Biomass energy and fuelwood can be renewed, but at present rates of use, woodlands and forests are being destroyed quicker than they can be replaced.

If we are to secure our future, it is important to find forms of energy that will not only continue to be available, but which will not damage the environment.

The perfect fuel?

There are many factors to consider when researching new sources of energy. A fuel may be renewable and non-damaging, but if it is too

A solar-powered public telephone in Australia. Solar power is particularly good for supplying localized electricity supplies such as this.

expensive or in a form that makes it difficult to use it is unlikely to provide a workable product. However, the development of new, renewable energy sources can also create jobs and save countries from having to buy expensive fossil fuels, giving them more control over their energy supplies.

FACT

Iceland is planning to become the world's first fossil-fuel-free economy and to create a hydrogen economy within 15–20 years (see pages 48–49). It will use its abundant hydroelectric power to provide energy to split water into hydrogen and oxygen.

The sun—the ultimate energy source

There are many ways to tap into the Sun's energy. The simplest is to use the heat to warm buildings directly, known as "passive solar heating." It has been estimated that, together with good design and insulation, this type of solar energy can provide 70 percent of a building's heating needs. Another type of design, known as "active solar heating," traps energy from the Sun's rays in specially designed collectors and pumps it around a building. This type of system can provide hot water as well as heating but it is more costly to install and needs more maintenance.

Electricity from the sun

Solar energy can be converted directly into electrical energy using photovoltaic, or PV cells (also called solar cells). You may have a calculator or a watch powered in this way. Large panels of

Large-scale solar dishes track the Sun so that the maximum amount of energy is collected. This group is in Australia.

solar cells mounted on the roofs of buildings, or on unused land or along highways, can provide large amounts of electricity. More ambitious projects involve huge computer-controlled mirrors that track the movement of the Sun and reflect its rays onto a central heat-collecting tower. One idea involves a vast, flat sheet of glass laid in the desert, causing heated air to be sucked up a tall "solar chimney," driving a turbine to generate electricity.

For and against

Solar heating (whether active or passive) is inexpensive and the technology needed is fairly simple. Pollution and land damage do not occur because the systems are built into existing structures and the energy source itself costs nothing. Solar cells are also capable of working in overcast conditions, although solar heaters do need direct sunlight. Solar cells are the ideal technology for providing electricity in isolated villages or in rural communities.

At present, solar cells are quite expensive to make and some pollutants are produced during their manufacture. As the technology improves and demand increases, the costs will fall. Racks of cells on a roof or in a backyard can look unsightly but less visible, thin, flexible rolls of cells are being developed to solve this. The technology is now also available to produce semi-transparent, solar-electric windows; they generate electricity and provide filtered light at the same time.

Toward the future

It has been estimated that solar cells could supply 15 to 17 percent of the world's electricity by 2010 (more than nuclear power supplies today). By 2050 the figure could be 30 percent (higher in some more developed countries). For some energy experts, the future looks sunny!

VIEWPOINTS

"If costed over 80 years, electricity from a solar chimney would cost a third as much as that from the two or three coal-fired power plants it would replace."
Wolf-Walter Stinnes, physicist

"You can make anything pay over 80 years."
David Infield, Center for Renewable Energy Systems Technology

FACT

People use more than 50 trillion kilowatt-hours of power per year. The same amount of energy reaches the Earth every 40 minutes in the form of solar radiation.

VIEWPOINTS

"The search for cheap, renewable sources of energy is losing momentum in many developed countries."
John Houghton, Intergovernmental Panel on Climate Change

"Offshore wind projects will make a substantial contribution to Britain's electricity needs."
John Battle, UK Energy Minister.

The way the wind blows

For hundreds of years, wind drove many of the mills that dotted the countryside. Now windmills are back, this time producing electricity rather than turning millstones. Today, there are over 30,000 wind turbines throughout the world, mostly grouped in clusters called wind farms. California alone has an amazing 17,000 machines. In Denmark, over 13 percent of electricity is produced by wind turbines and this figure is expected to rise to 15–16 percent by 2002.

No wind shortage

Wind power is a practically unlimited source of energy. The global potential of wind power is five times the present electricity use. Wind farms can be built quickly and expanded as needed and they can produce fairly high yields of energy with little or no pollution. The energy source itself is free and the land on which the wind turbines are built can be used for grazing animals or other agriculture.

Get that windmill off my land!

To work efficiently, wind turbines need steady winds and back-up electricity from another energy source when the wind dies down (electricity that could come from a second renewable source).

But the main objections to wind power are "visual pollution" and

In this small-scale project in South America, a windmill powers the pump that brings essential water supplies for irrigation and livestock.

noise. Although modern turbines can be quite beautiful, huge groups of them spread across the countryside can spoil the view. Winds blow more strongly in high upland areas, and turbines are often located in these regions where they are even more visible.

The noise of hundreds of turning turbines is considerable and can be a constant source of annoyance. Large wind farms can also interfere with the flight patterns of migrating birds. In some cases, birds of prey that hunt in the kind of terrain that is ideal for wind farms, have been killed by flying into the blades.

FACT

More efficient designs of wind turbines, with lower maintenance needs, may make it possible to place wind farms in very remote areas, such as high up in mountains, where they cannot be seen or heard.

A windy future?

One possible solution to visual and noise pollution is to build wind farms out at sea. But this will be far more expensive and technologically difficult. Even so, wind power experts predict that, by the middle of this century, wind power could supply more than 10 percent of the world's electricity.

A large bank of windmills dominates the landscape at this wind farm in California. Many people complain of the "visual pollution" this causes.

Ocean waves contain unimaginable amounts of energy, but harnessing it effectively is a different matter.

Harnessing the sea

The sea is a huge and free source of potential energy. Even if we could trap only a tiny percentage of its power, we would have a pollution-free and renewable source of energy for as long as the world turns and the wind blows.

As the world turns, the gravitational pull of the Moon draws the world's oceans toward it, creating the twice-daily tides. The potential energy contained in this great movement of water is enormous. Attempts have been made to harness the power by building barrages across bays and estuaries, where the tidal effects are most concentrated. This has been tried commercially at the mouth of the Rance River in Brittany, France.

However, estuaries are often important feeding grounds for seabirds, and barrages change the pattern of tidal flow, flooding the mud flats on which they feed. For this reason, suitable sites are scarce, while construction costs are high.

FACT

Most experts predict that tidal power will only make a tiny contribution to world electricity supplies.

Wave power?

Waves are created mainly by wind blowing over vast expanses of ocean and they contain a huge amount of energy. In recent years, a number of ingenious designs have been developed to try to capture this energy. The best known is "Salter's duck" named after its inventor Stephen Salter. Large floating vanes rock back and forth on a central shaft, compressing liquid that drives a turbine. In another design called an OWC (Oscillating Water Column), large pistons are

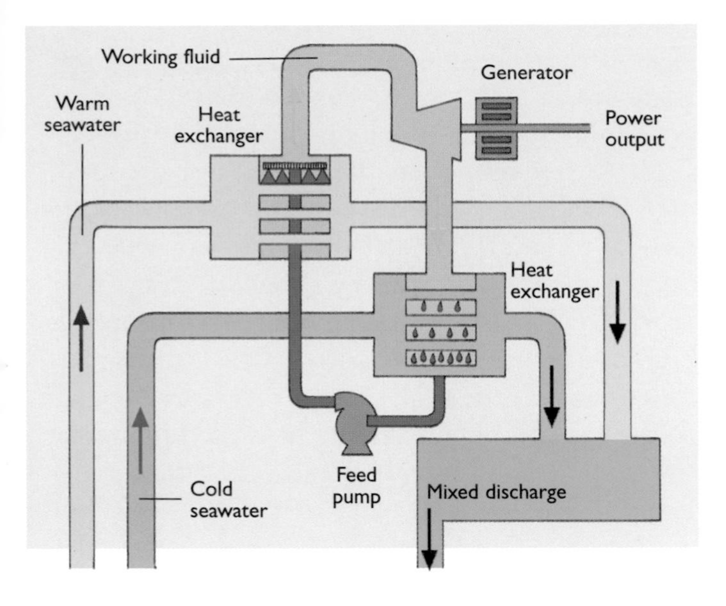

In this experimental system, the difference in temperature between warm surface water and colder, deep water is used to evaporate and then condense ammonia via heat exchangers. The ammonia vapor drives a turbine to produce electricity; the seawater is returned to the sea.

driven up and down by the waves, compressing air that connects to a turbine to produce electricity.

In the past, wave power generators have mainly met with failure. Now, 15 more robust, better designed machines are planned across the globe, nine of them around Europe's coasts where the Atlantic pounds the shores. Even so, the future of wave power is uncertain. Much will depend on how the present series of generators performs over the next few years.

Sea radiators

Japan and the U.S. have been exploring the possibility of exploiting the difference in temperature between the cold depths and the sun-warmed surface water of tropical oceans. The temperature difference is used to evaporate liquid ammonia which then drives a turbine. Although energy from this source would be almost unlimited, the costs of what is known as ocean thermal energy conversion (OTEC) means it is unlikely to compete well with other energy alternatives.

VIEWPOINTS

"We could generate half of Britain's electricity requirements from waves by 2040—I am convinced we can do it."
Gordon Sinclair, British task force on wave power

"I am too battle-scarred to be over optimistic about the future."
Stephen Salter, inventor of "the duck," University of Edinburgh

FACT

About 20 countries worldwide extract energy from geothermal sites. The U.S. produces 44% of all the geothermal electricity generated in the world and Iceland, Japan, New Zealand, and Indonesia also use it.

Energy from underground

The Earth itself is one huge energy generator. Radioactive rocks containing uranium-237 and potassium-40 decay, producing huge quantities of heat. At the same time, the enormous pressures created deep beneath the Earth's crust cause rocks to heat to unimaginable temperatures. This is known as geothermal energy. Some of this geothermal energy heats the water trapped inside porous rock in the Earth's crust, which turns into super-heated water and steam. If it is close enough to the surface, bore holes can be drilled to extract the steam or hot water which can be used for heating or to generate electricity.

For and against

Geothermal energy provides a reasonably inexpensive and very reliable source of energy, but only at or near reliable geothermal sites. It also produces 96 percent less carbon dioxide emission than fossil fuels. However, it is available only in certain parts of the world, and supplies of steam or

Bathers in this thermal pool in Iceland enjoy the benefits of underground geothermal energy. This type of energy is of major importance to countries such as Iceland and New Zealand, which are situated near active regions of the Earth's crust.

This greenhouse in Iceland is heated by geothermal energy.

hot water could be quickly depleted if the rates of extraction are greater than the rates at which they build up. The older types of geothermal plant also produce a moderate amount of water pollution.

With new technologies, most of these drawbacks are likely to be eliminated. In new systems, the energy in the steam or hot water is transferred to another fluid so that the original fluid can be re-injected into the ground. This eliminates most of the environmental problems of the older systems since the water remains in the rocks.

It's all in the rocks

The main barrier to greater use of geothermal energy is cost. Only where sites are concentrated and easy to access is it worth extracting the energy. However, with the current development of new systems for reaching hot rocks deep inside the Earth, geothermal energy may become an important energy source for the second half of this century.

FACT

Japan is planning to use geothermal heat from 650 feet beneath the ground to keep roads ice-free. The energy will be brought to the surface and used to heat antifreeze which will be pumped around a network of pipes 4 inches below the road.

Scientists processing liquid hydrogen to develop cleaner ways of producing energy.

Hydrogen—the miracle fuel?

"We are at the peak of the oil age but the beginning of the hydrogen age."
Herman Kuipers, Global Solutions (Shell Oil)

In many ways, hydrogen gas is the ideal fuel of the future. The reason is simple: when hydrogen burns in air it produces only one by-product—water. There is no pollution of any kind.

However, hydrogen gas has to be produced. One method is to produce it from compounds called hydrocarbons, made up of carbon and hydrogen. Since these are the main ingredients of gasoline and natural gas, carbon dioxide is also produced. Obtaining hydrogen from water is better since only harmless oxygen is released.

Something for nothing

Of course, we can never have something for nothing. Splitting the water into hydrogen and oxygen uses energy which, if it comes from fossil fuels or nuclear power, is of no benefit. But if the energy used comes from a renewable source, such as

VIEWPOINTS

"Hydrogen is a tehnological dead-end because of the high energy needs of producing and storing the hydrogen."
Greenpeace, Germany

"The big attraction of hydrogen is that it will help solve urban air quality problems."
Friends of the Earth, UK

solar energy, and can be produced cheaply enough, then we have the real possibility of developing the ultimate non-polluting, renewable fuel.

Clean running cars?

It is possible that hydrogen could also be used to power vehicles, making air pollution a thing of the past. Scientists are developing systems called fuel cells, in which the hydrogen is combined with oxygen by a process called electrolysis, to produce electricity. Some prototype vehicles have already been produced, including a car that can travel 250 miles on 10.5 gallons (400 km on 40 liters) of methanol (from which the hydrogen is produced). In addition, a number of buses and even a submarine have been designed to run on fuel cells. It is also possible that power plants may eventually produce electricity from fuel cells based on hydrogen.

DEBATE

Is there a danger that unforeseen problems could arise with renewable energy sources as they did with nuclear power? Should governments invest more in alternative energies? Who decides a country's "energy policy" and how much of a choice do we really have in determining our future energy sources?

This Chrysler car has a hybrid power unit; the rear wheels are powered by a gas engine and the front wheels are powered by electricity produced from hydrogen.

GETTING MORE FROM LESS

FACT

Building standards (affecting house design and insulation) in Sweden mean that energy-use for heating is far lower than in other countries, despite Sweden's colder climate. Similar standards used in Milton Keynes in the UK have reduced heating bills by 40% with only a 1% increase in building costs.

Delivering wall insulation to a newly built house in Denmark, in line with energy-saving building regulations.

Is Your Fridge Efficient?

There are two ways to tackle the world's energy problems. One is to produce better, cleaner fuels, the other is to use less energy. Conserving energy and using it more efficiently not only means less pollution and less damage to the environment, it also saves money.

Throwing it all away

In the U.S., about 80 percent of commercial (generated) energy used is wasted. About half is lost during energy conversions (such as when producing electricity from oil—see page 51). The other half is wasted unnecessarily, for example in "gas-guzzling" motor vehicles and industrial processes, and in poorly designed and badly insulated buildings.

Increasing energy efficiency

Energy efficiency is the percentage of total energy used that actually does useful work. For example a normal light bulb has an efficiency of only about 5 percent which means that 95 percent of the energy used is wasted, mostly as heat. A car engine has an efficiency of around 10 percent, and an energy-saving or fluorescent bulb

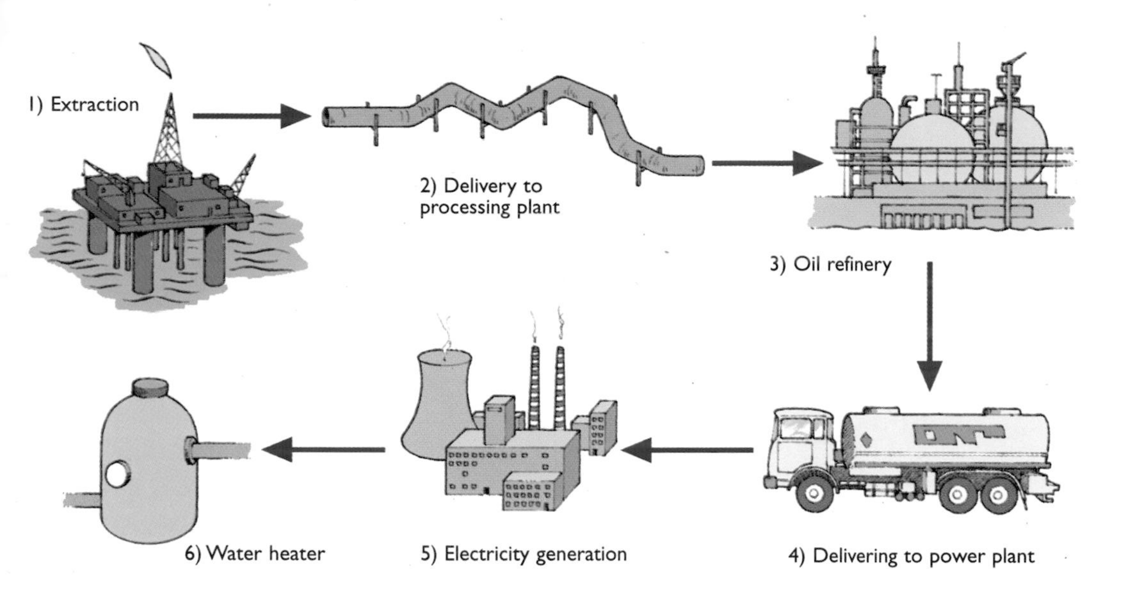

around 22 percent (about the same as the human body). A steam-driven turbine (producing electricity) has an efficiency of about 45 percent. Fuel cells (see page 49) come high in the efficiency count at around 60 percent.

Using electricity to heat water is a waste of energy. At each stage of the process, an energy conversion takes place, during which some of the original energy is lost as heat or some other form of unusable energy. As a result, over 85 percent of the energy in the oil is wasted. Heating the water using solar energy, or even directly by gas, would be more efficient.

One of the factors affecting energy efficiency is the number of steps, or energy conversions, needed to get from the original energy source to the end use.

The energy loss for the series of conversions shown above is over 85 percent (i.e., about 14 percent energy efficiency). For the same heating effect to be achieved by passive solar heating (one conversion only, see pages 40-41), the energy loss is only about 10 per cent (i.e., 90 percent energy efficiency).

The right energy for the right job

Electricity is sometimes called "high-quality" energy. It is useful for televisions, for example, or high-speed machinery, but it is wasteful when you may only want to heat air or water to fairly low temperatures.

VIEWPOINT

"So long as energy is cheap, people are more likely to waste it and not make any investments in improving energy efficiency."

G. Tyler Miller, Professor of Human Ecology

FACT

More energy-efficient appliances provide financial as well as energy savings. Unfortunately it is often those who would most benefit, such as the less poor, the elderly, and those in poor housing, who can least afford them.

Reducing waste

Industry is already much more energy efficient than it was ten or twenty years ago. With rising energy costs (particularly costs of oil) it makes sense for producers and manufacturers to cut their fuel bills.

Many electric motors in machinery are run at full-power all the time rather than at the speed that is actually needed. Now, computer-controlled systems can be used to turn off lighting and equipment where they are not required and make adjustments when production levels are low. Many industries produce large amounts of waste heat that can be used to produce steam to drive turbines and generate electricity. Alternatively, the waste heat from generating electricity can be used for localized heating, a system known as co-generation.

The control room of a power plant. During times of low production, slowing down or turning off machinery prevents energy being wasted.

Energy can also be saved by switching to high-efficiency lighting and making more use of natural lighting. Recycling materials and making products that last longer or are easier to recycle also saves energy, reduces pollution and the use of other resources.

Greener homes

It has been calculated that the heat loss in the average home is equivalent to having a large, window-sized hole in the side of the house. Besides insulating buildings, energy loss can be reduced by using the most energy-efficient appliances available. Although initial costs may be higher, there is often a saving when operating costs (known as life-cycle costs) are taken into account.

DEBATE

Is wearing another sweater really going to make a difference? Greater energy efficiency and waste reduction can save energy, but there is a limit to what it can achieve. Do we need to think more deeply about the way we think about energy, rather than coming up with yet more technological "fixes"?

The energy-saving measures in this Danish geodesic ecohouse include the central heating "tower," insulated walls and large solar windows.

Greener living

Changes in our lifestyle can achieve the greatest savings. Simple things such as switching off lights, putting on a sweater rather than the heat, walking or bicycling and using public transportation whenever possible, can make a difference.

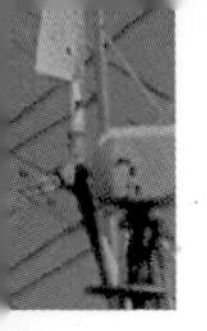

ENERGY FOR THE FUTURE

VIEWPOINT

"If the United States wants to save a lot of oil and money and increase national security, there are two simple ways to do it: stop driving gashogs and stop living in energy sieves!"
Amory Lovins, energy expert

Development Versus the Environment?

We have to strike a balance between allowing countries to develop economically (which may mean using more energy) and protecting the environment from the effects of that energy use. In particular, it is important that developing countries gain the benefits of further development so that people worldwide can enjoy a decent standard of living. But how can this be achieved without irreparable damage to the environment?

Agreement in Kyoto

In December 1997, representatives from 167 countries met in Kyoto, Japan, to discuss how that question could be tackled. Five years earlier, they had agreed at the biggest on possible actions environmental conference ever held, the Earth Summit in Rio de Janeiro, Brazil. Now it was time to make that commitment work.

Ensuring that everyone benefits from development, without destroying the environment in the process, is the greatest challenge of this century. These children are playing on a beach at Lake Michigan, where industries from the nearby city of Gary, Indiana, pollutants into the lake.

The Kyoto meeting had one main strategy—to place limits on the richer countries, emissions of greenhouse gases such as carbon dioxide and methane. Each developed or industrialized nation agreed on a specific reduction, with the overall aim of cutting emissions of greenhouse gases by about 5 percent by the year 2010. There were no specific limits placed on developing countries since it was felt they should be allowed to develop without such restrictions.

VIEWPOINT

"The international climate negotiations that have taken place would allow the developed countries to escape curbs on their own emissions in return for helping us. This is unfortunate."

Nakibae Teuatabo, Climate change co-ordinator, Kiribati, South Pacific Islands

Fair swap?

A proposal called the Clean Development Mechanism (CDM) allows developed countries to help poorer countries develop cleaner energy technologies in return for allowing the rich countries to relax their own CO_2 reductions. Some, however, feel this is not a fair swap.

Difficult issues

These are difficult and complex issues. We do not yet know how successful the Kyoto agreement (or "Protocol") will be. Even if all the targets in reducing emissions are met, far greater reductions will be needed to stop the build-up of greenhouse gases in the atmosphere. What it will show, however, is that countries can work together to achieve change.

This graph shows energy consumption over the past 30 years and the likely future consumption of the major economic areas of the world. The rich developed countries (Europe, U.S., and Japan) have the highest overall consumption, but the fastest growth rate is occurring in Asia. What do you notice about energy consumption in Africa? How would you explain this? Which economic area shows the least growth? Why do you think this is?

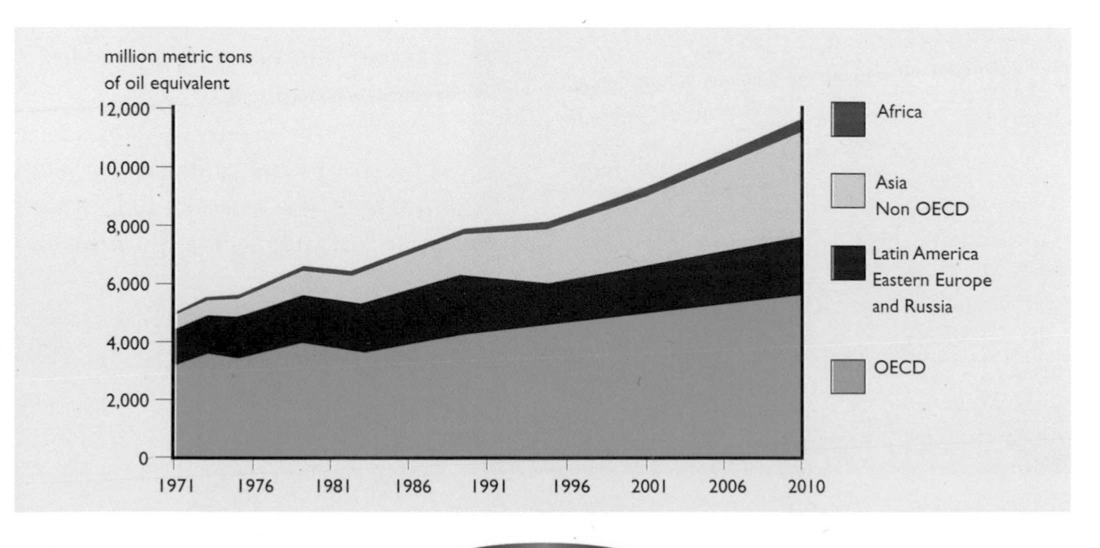

FACT

The U.S. is among several industrialized countries that have failed to live up to their Earth Summit commitment to reduce emissions of greenhouse gases to 1990 levels by 2000.

Where do we go from here?

This book has looked at a number of important issues about our use of energy in the twenty-first century.

The first concerns our continuing use of fossil fuels. It is clear that fossil fuels will go on being our most important source of energy for several decades to come, perhaps longer. The major question is—how much further damage to our global environment will this use do, and will there come a point when it becomes irreversible? Another crucial question is—what will we use instead when fossil fuel supplies run out?

The second issue is about poverty. People in developed countries have achieved a high standard of living by exploiting energy sources massively and wastefully, and with very damaging effects. Now other, less-developed economies want their turn. The question is—How can the developing countries achieve better conditions for their people, without serious (or possibly catastrophic) consequences for the environment?

The third issue is about the way in which the world develops its energy strategies for the future.

The environmental organization Greenpeace gets its message across in a unique way using this coal-fired power plant in Germany. Global warming is arguably the most serious issue the world faces today.

There are two questions here. The first is—How can we reduce our demand for energy as populations continue to rise? The second is—in what direction should we take in our search for less damaging, renewable sources of energy?

A question of politics?

Most of the renewable sources of energy, such as the Sun, water, wind, and hydrogen, are "free at source." This means that individual countries and even communities will have more control over their own energy generation and use. They will no longer be dependent on big energy suppliers, such as the oil-rich countries or international energy companies, to supply their needs. This "independence from oil" may be important in determining how countries develop in the future, both politically and socially.

Energy is not just about the environment—it is about the kind of society we are moving toward.

VIEWPOINTS

"Developing countries should be allowed to improve their economies first; then they will automatically clean up their environments. This is what happened in the developed countries."
Theory based on the ideas of Simon Kuznets, Professor of Economics

"The environmental damage done by the rich countries was due to ignorance and primitive technologies. Poor countries today have not got the same excuses. They should not make the same mistake."
Paul Elkins, Forum for the Future

Belgian truck drivers protesting against the high price of fuel. But are high prices necessary in order to reduce our use of fossil fuels?

This mobile solar-powered kitchen was developed by the Greenpeace organization to demonstrate the potential for clean, renewable energy.

A long way to go?

This book began by saying that all living things, including humans, need energy in order to survive. But *how much* energy we need is a different matter. This book also suggests that we use energy wastefully and often without much thought for the global consequences.

Until realistic and economic alternatives are developed, we will go on being dependent on fossil fuels (and possibly even nuclear power) for some time to come. It makes sense, therefore, to use energy as sparingly as possible.

Doing the simple things

There are many ways in which we can save energy in everyday life. These include switching off lights, TVs, and computers while not in use, using less hot water and heating, and finding alternatives to the car, such as public transportation, bicycling, walking, or even roller blading! These benefit the environment, save money, and are easy to do.

The low-energy school

Saving energy is not just something to do at home. Schools are often great energy wasters! There are many programs that involve pupils, teachers, and parents in school energy-saving plans. Since schools are responsible for their own energy bills, the savings can be easily measured and can be re-invested in other improvements.

Thinking energy

Even as alternatives to present energy sources become available, we should still use energy with much thought and care. As we have seen throughout this book, energy never comes free, and never comes without some kind of cost to the environment, however great or small. To put it simply, our attitudes toward energy are really about our attitudes toward the Earth.

DEBATE

Are we really facing an "energy crisis"? If so, what are the possible consequences of that crisis (a) for ourselves (b) for others (c) for the environment? Do you think those in power take the issue seriously enough? If not, what should they be doing? Should we adopt the so-called "precautionary principle" and take tough, energy-saving measures *now*, even if we are not yet certain of the effects of our current energy use?

There is no such thing as free energy. We have to make wise energy choices now or we will end up, like these people, with very little room to maneuver in the future.

GLOSSARY

acid rain rain or snowfall polluted by harmful emissions in the atmosphere.

biomass plant and animal material used as fuel.

Clean Development Mechanism 1997 agreement encouraging rich nations to help poor nations develop cleaner energy production methods. In return, the rich nations are allowed higher levels of greenhouse gas emissions than those imposed at the Earth Summit in 1992.

developed countries rich industrial nations, e.g., U.S., most of Europe and Japan.

developing countries poor, non-industrialized countries, such as many in Africa and Asia.

economic development way in which a country gains wealth, e.g., by becoming more industrialized.

electrolysis process used to split chemical compounds into smaller elements, used to develop fuel cells from hydrogen and oxygen.

"elsewhere pollution" emissions from an indirect energy source, e.g., the source for recharging non-polluting fuel cells.

emissions waste gases (e.g., carbon or sulfur dioxide) from transportation or industry.

energy efficiency percentage of total energy that does useful work (e.g., a normal lightbulb is 5% efficient).

fossil fuels coal, oil and gas.

generator machine that converts mechanical energy into electrical energy.

greenhouse gases gases from burning fossil fuels that warm up the atmosphere, causing global warming.

insulation material, or process, that reduces the rate of heat loss.

isotopes heavier versions of atoms (e.g., plutonium-239) which split when they decay to release radioactive energy.

joule unit that measures energy.

life-cycle costs total costs of a product in its entire life-cycle, including extraction, transportation, manufacture, use, and disposal.

potential energy energy, such as water held by a dam, which must be converted into another form to be used.

reactor the part of a nuclear power plant in which the splitting of atoms occurs.

renewable energy source that won't run out (e.g., wind, water, Sun) and is less polluting than non-renewable sources (e.g., fossil fuels).

solar (PV) cells panels which convert sunlight's energy into electrical energy.

turbine structure fixed to an electrical generator turned by water, steam, or wind.

water cycle natural process by which water evaporates from seas and rivers into the atmosphere then falls to Earth as rain or snow.

watt unit that measures power.

MORE INFORMATION

Conway, Lorraine. *Energy*. (Superific Science series). Good Apple, 1985.

Grimshaw, Caroline. *Energy*. (Invisible Journeys series). World Books, 1998.

Gutnik, Martin J., and Browne-Gutnik, Natalie. *Projects That Explore Energy*. (Investigate! series). Millbrook Press, 1994.

Morgan, Sally, and Morgan, Adrian. *Using Energy*. (Designs In Science series). Facts on File, 1993.

Spurgeon, R. *Energy & Power*. (Science & Experiments series). EDC, 1999.

Wolven, Doug. *Energy*. (Investigations in Science series). Creat Teach Press, 1996.

USEFUL ADDRESSES

Enviromental Protection Agency Home Page:
"…to protect human health and to safeguard the natural environment…"
http://www.epa.gov

American Petroleum Institute Home Page:
API is the major national trade association representing the entire petroleum industry: exploration and production, transportation, refining, and marketing.
http://www.api.org

U.S. Department of Energy:
The Department of Energy's mission is to
foster a secure and reliable energy system that is environmentally and economically sustainable, to clean up our own facilities and to support continued United States leadership in science and technology.
http://www.energy.gov

Natural Resources Canada:
A federal government department specializing in energy, minerals and metals, forests and earth sciences.
http://www.NRCan-RNCan.gc.ca/homepage/index.html.

INDEX

Numbers in **bold** refer to illustrations.

© 2001 White-Thomson Publishing Ltd